CONCEPTS AND TECHNIQUES IN MODERN GEOGRAPHY No. 18

QUESTIONNAIRES AND INTERVIEWS IN GEOGRAPHICAL RESEARCH

by

C.J. Dixon and Bridget Leach

(City of London Polytechnic and Thames Polytechnic)

CONTENTS

I DATA COLLECTION AND RESEARCH DESIGN

(i) Introduction

Any research project sets out to answer specific questions or to illuminate a particular subject area. In commercial or market research, the problem is clearly defined by the client, but at the outset of academic research, the aims may be less rigid. However, the first step in designing a research project must be to set out a definite central problem or field of investigation.

A project cannot be designed if its objective is vague. We cannot set out to 'find out about shopping', for example, although if we narrowed down our interest to the problems of pensioners and shopping, we would have begun to define the population of interest and started on the formation of hypotheses. These might be as imprecise as speculation about travel difficulties and prices paid, but unless such concerns were set out, it would be impossible to work out the kind of topics to be covered by the research, let alone the phrasing of questions.

It is not necessary to start with formally worded hypotheses about relationships between phenomena, but even the smallest scale project must have, at an early stage in its design, definite boundaries to its field of study. Focusing and refining of hypotheses and subject matter can only then begin.

Preliminary investigation of literature and secondary material (directories, maps, government and local authority records and statistics) must be made next, to find out if the information needed already exists in a usable form. If it does not, appropriate data collection methods can be devised.

One way of obtaining primary data is to conduct a survey, to ask questions of a number of respondents. If the data are to be quantifiable and the findings generalised, the questions must be standardised and the respondents chosen in a scientific manner. It might appear that anyone could design and execute a survey, because we are all used to asking questions and obtaining answers, but although most questions will be answered, it is not always possible to know that the respondent has understood the question intended. It is also possible to ask questions which fail to illuminate the problem under investigation. With careful planning, pilot testing and revising, these problems can be minimised, but it is also vital to select a project which is not over-ambitious, and which can be tackled with the time and resources available.

Surveys take two main forms, for questions can be put by an interviewer and the answers recorded on an interview schedule which sets out the questions and provides room for the answers, or a questionnaire can be filled in by the respondents themselves. This handbook outlines the use of these two specialised techniques, interview and questionnaire surveys.

(ii) Structured and unstructured techniques

A good journalist might find out more than a researcher with a bad questionnaire, but he would have greater difficulty in providing evidence for his conclusions. There is no way of knowing the representativeness of the insights gained from a small number of people, whereas the numbers and percentages obtained from a probability sample can be said to represent the population being studied with a calculable probability of accuracy. Sampling methods are discussed in CATMOG 17 (Dixon and Leach, 1978).

However, unstructured interviewing techniques may yield valuable understanding, and before explaining the detailed steps of survey work, it is worth considering their use. Frequently, structured and unstructured techniques are not alternatives, but complementary. A project may benefit from informal discussions with a small number of respondents at the beginning, before a formal survey with a standardised interview or questionnaire. For example, perhaps we might want to find out how people find their way around a new area. We could design a survey which asked a sample of recent movers if they bought a street map, or how they found specific facilities such as a doctor, but we could not immediately be sure if we were asking about the right sort of things. Group discussions, or long interviews, might reveal more about the ways of exploring unfamiliar places; from these it would be possible to develop questions for a quantifiable survey. After the survey itself has been analysed, many loose ends, contradictions and problems can be illuminated by longer, unstructured interviews.

In a survey, each respondent must be presented with the same question, with nothing omitted or added which could change his interpretation of what is required. A good interview should flow in a way approaching conversation, but should actually be carefully standardised and controlled by the interviewer.

In some small surveys there is only a small amount of data needed from each respondent, but additional detailed information may be obtained from a number of the respondents, and it may be useful to discuss the issues raised by them at greater length. Here a semi-structured conversation, using a simple interview guide which lists topics to be covered, may be adequate to obtain comparable data. The exact order of the questions would not be formalised, but the guide would be there to ensure that nothing was left out. This method would be most appropriate if a sample survey was not being undertaken, for example where the population of interest was small, such as the industrialists on a trading estate or the farmers in a limited area.

Recording information from unstructured interviews presents problems. A good memory, (and immediate writing up to minimise the rapid 'memory decay'), shorthand, tape recorders, or several research workers taking notes at the same time all have specific applications. After the first few minutes, a tape recorder seems not to be off-putting, but the researcher probably lacks the facilities for transcribing the interview, and is merely postponing the problem of coding and analysing the information obtained.

Unstructured techniques are not discussed here in detail, but if a researcher feels that they are appropriate to his or her project, the SCPR (197: manual can be recommended.

Both structured and unstructured techniques depend on the skill and integrity of the researcher. When reporting either, the exact methods adopted and the questions asked should be described, so that the reader can assess the limitations and strengths of a piece of research.

A formal survey is not something which should be undertaken without considering alternative data sources and research methods, but with careful design information can be collected which could not be obtained any other way.

(iii) Designing a survey

The design and execution of a survey must proceed logically, and Table 1 sets out the general sequence of operations to be carried out. At an early stage the whole plan from the initial formulation of ideas and hypotheses through to analysis and writing up should be mapped out. Sponsors or a supervising department may need a time and cost budget, and estimates can be made for each stage.

The scope of a project will depend on resources of time, cost and manpower, and decisions about the survey area and the population of interest will have to be made and modified in the context of these resources. The more variable a population, the larger the sample needed to produce accurate estimates of its characteristics (Dixon and Leach, 1978). A national survey of adults in Britain might need 2000 respondents, but it is possible to obtain accurate results from small surveys which are less ambitious in their coverage.

To increase the size of the sample, interviews could be shorter, allowing more to be conducted in the same time, but with limited resources it would be more effective to reduce the variability in the range of answers obtained, which enables a smaller sample to produce results which are of an acceptable level of precision (Dixon and Leach, 1978). This could be done by asking only about attributes which did not vary greatly through the population, or by confining the study to one age group or household type or setting spatial limits to the study, on the assumption that the respondents would therefore be relatively homogeneous.

Once the population and broad coverage of the questions have been fixed decisions can be made about the most appropriate survey methods (chapter II). Then it may be found that resources are inadequate to deal with the survey unless redefinitions are made. Research work is usually a compromise between initial ambitions and feasible operations but it is better to carry out a good survey of a small topic than to produce inconclusive work on a larger one.

Each question must be designed to obtain information related to the topic, and not information which appears 'interesting' or which may prove relevant. The most carefully conceived project may leave data uncollected which will be later regretted, and ask redundant questions which throw no light on the study, but these problems can be minimised by careful planning.

The respondents, too, should be able to see the connections between questions and the central concerns of the study, which will have to be explained to them in order to obtain their co-operation.

Table 1: The Main Sequence of Operations in a Survey

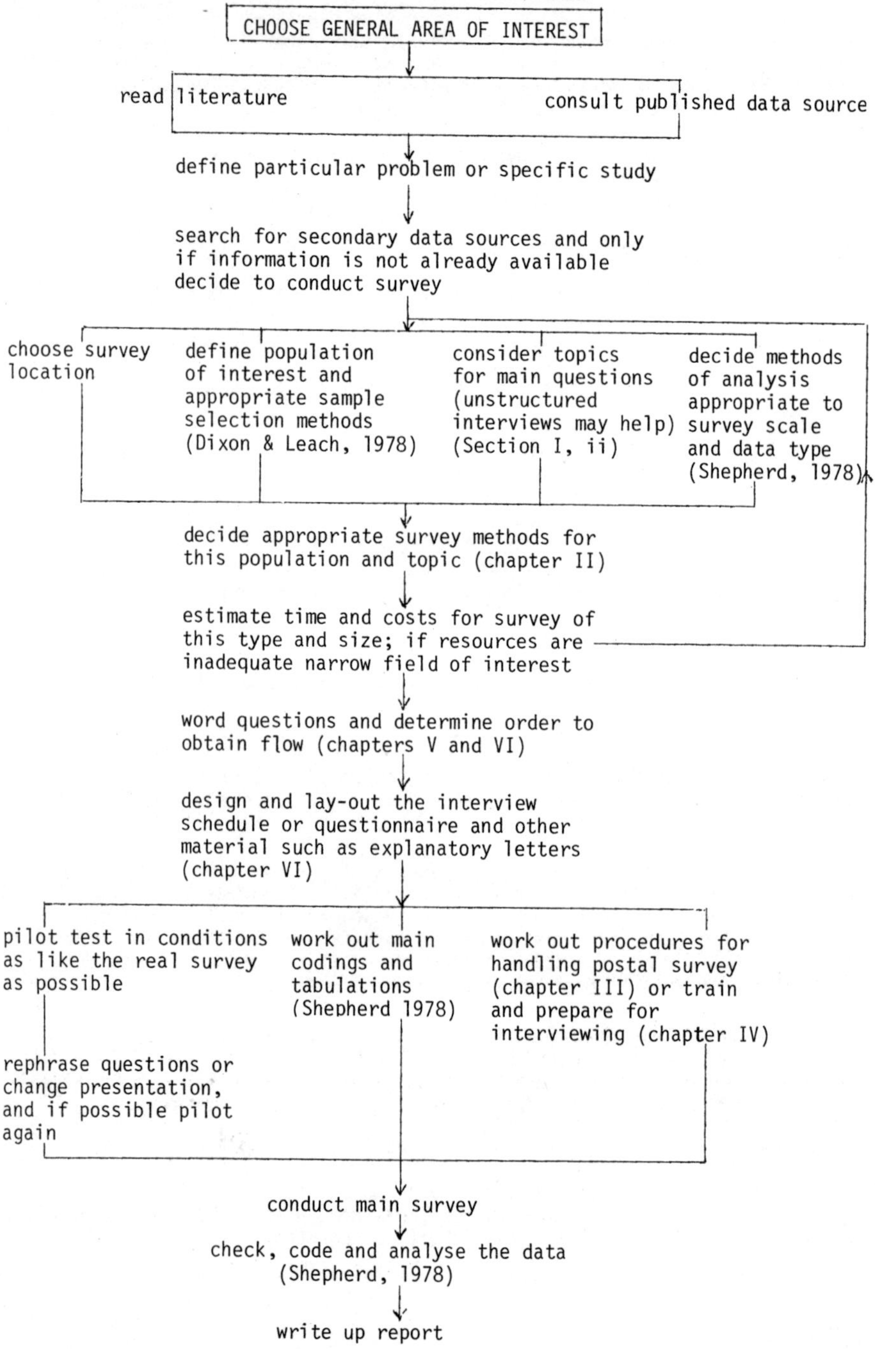

Even at the earliest stages of designing questions, consideration must be given to how coding and analysis are to be undertaken (Shepherd, 1978). Samples of less than a hundred cases may be amenable to hand calculation, but larger surveys will need computer analysis, and an inexperienced researcher will need advice on this from the outset.

Time spent delimiting the research, consulting the literature, and designing and pilot testing the format of a questionnaire or schedule will not be wasted. These tasks cannot be carried out hastily, even for the most straightforward topic with which the researcher feels confidently familiar. Good planning will actually speed the operations, particularly if thought is given from the beginning to the analysis itself.

II CHOOSING A TECHNIQUE: INTERVIEWS OR SELF-ADMINISTERED QUESTIONNAIRES

(i) The choice

There are basic differences in the approaches adopted in questionnaires which the respondent fills in himself and interviews where the answers are recorded by a field-worker familiar with the schedule. Much of what will be said in later sections of this handbook applies to both forms of asking questions, but details of the applications will differ.

At an early stage in research design, the researcher has to decide which technique is best suited to the data he wants to collect, the target population his sample sets out to cover and his resources. This section discusses the factors he has to take into consideration, summarised in Table 2.

(ii) Content and coverage

Questionnaires need to have a simple structure, and are unsuitable for the type of question needing detailed instructions. A respondent on his own may well read the whole form before starting to fill it in, and there is no possibility of using one question to check on another, or of obtaining spontaneous replies. The respondent may well ponder his reply, and choose one which presents a particular image, even subconsciously, or he may be helped by others. An interviewer could often gauge this, and make marginal notes which would explain the circumstances (so that in extreme cases the schedule, or a particular question, could be discarded).

On self-administered questionnaires, questions may be left blank, misunderstood or incorrectly filled in. Where unsatisfactory returns are widespread, it suggests that the question(s) or technique were wrong for either the topic or the respondents; pilot testing should discover this.

Respondents may be naturally reluctant to discuss subjects they consider personal, which may include behaviour, attitudes, income and details of their families. Although there will always be some who cannot be persuaded to revea sensitive information, even with guarantees of confidentiality, most respondents will be helpful if the reason for the question is clear, and the aims of the research seem reasonable.

It may be argued that the presence of an interviewer can give the necessary reassurance and explanation, but conversely, the impersonal nature of a questionnaire may seem more suitable for some topics and some people. Both a good questionnaire and a well-designed interview should be able to elicit sensitive information, and the choice between the two methods will depend on the amount of detail required. Where there are a large number of supplementary questions which depend on the first answer, an interview may be preferable because otherwise the questionnaire would have to be very long even though not all questions were applicable to each respondent. If the instructions would have to be very complex to enable a sequence of questions to be understood, then it is best to rely on an interview rather than a self-administered form.

Geographers will rarely want to find out very sensitive information, but incomes, rents, and so on, should only be asked where they are strictly necessary to a particular survey, and a reason for asking about them should be given if it is not obvious.

Where information is needed from a number of members of a family, household or business, or has to be checked from documents or records, a questionnaire filled in at the respondents' convenience can be used.

There is no guarantee with a self-administered questionnaire that the person to whom it was addressed was the one who actually filled it in, even if a covering letter explained the importance of a particular person doing so. In view of the frequency with which husbands and wives in particular pass the form from one to the other, it might be sensible to restrict the use of postal questionnaires to surveys of households as a whole, or to businesses where the information sought could be supplied by one of several people.

(iii) Population and sampling frame

Most surveys cover a sample of the population that is being studied (Dixon and Leach, 1978). Samples need to be identified, and the way in which this is done may determine whether an interview or questionnaire survey is to be used. A personal interview may be dictated when the sample has to be contacted in the street, or consists of people using a particular facility.

Frequently there is a list which contains names and addresses of the target population from which a sample can be drawn. Without such a sampling frame, a postal questionnaire could not be used, although if the units in the population (households, businesses) could be identified on the ground, a form delivered and collected would be possible.

In areas of high population turnover, postal questionnaires are unlikely to achieve a high response, as they would either have to be addressed impersonally (for example to 'the occupier') or would stand a good chance of being directed to someone no longer there. Areas with better educated respondents might produce higher response rates, but if the survey was particularly concerned to include members of the population who were less well-educated, immigrants with possible language difficulties or very busy people unlikely to fill in a form without personal encouragement, the presence of an interviewer might be essential.

(iv) Response

Whether interviews are conducted in the street or respondents are approached by name in their own homes, we can expect that most people can be persuaded to be interviewed, provided that not too much time is demanded of them. Good interviewers can often turn an initially unfavourable reaction into an agreement to participate. Response rates are usually over 70 per cent.

Postal surveys have produced response rates in excess of 90 per cent, usually if the respondents are particularly interested or concerned with the subject of the study, but responses as low as 10 per cent are encountered. The larger the non-response, the less meaningful the conclusions - a factor which is exacerbated by the tendency for certain types of respondent to select themselves. A recently reported study of attitudes to education conducted by a local Conservative party obtained 13 per cent replies, mostly from parents; there is no way of knowing whether those who replied were sympathetic to the sponsors; they might be those with the strongest feelings about education. Either way no conclusions could be drawn. A study in which breweries were sent questionnaires about production of a particular sort of beer produced a 41 per cent response, and there were suggestions that the non-respondents were very different in their policies and practices from those who did reply.

In a postal survey, later respondents can be very different from early ones; they may be more like those who did not reply at all, although this cannot be assumed. Large scale surveys sometimes send out postcards with brief questions to those who have not returned the original form after a couple of reminders, in the hope of finding out if there is a major category of the population excluded altogether. A more costly and time-consuming alternative is to try to interview a sub-sample of the non-respondents. These techniques are like shutting the door after the horse has bolted, since careful planning could have produced a higher response or indicated the unsuitability of the technique. Probably any survey would do well, though, to compare the characteristics of the respondents with any known characteristics of the target population they are supposed to represent. For example, in a general survey of the residents of an area, key variables (age, sex, family size, and so on) could be compared with the most recent Census figures for the same area. If the survey area did not correspond exactly to the areas used by the Census, data for the approximate area would give at least some indication of accuracy.

(v) Scale

Where the target population is scattered over a large area, as might be the case with firms of a particular type, a postal questionnaire will save the time and cost of travelling to each one in the sample. To minimise travel costs, large commercial interview surveys choose samples that are clustered spatially so that each field-worker has to interview over a part of the area. The disadvantages of cluster samples are discussed in CATMOG 17 (Dixon and Leach, 1978).

Many small scale projects will be restricted to a small locality, and therefore interviewing or delivery and collection would both be feasible alternatives to postal surveys.

With a self-administered questionnaire, a large number of respondents can be contacted by a lone researcher. A single field-worker can rarely conduct enough interviews before beginning unconsciously to bias answers in line with expectations, and before boredom leads to carelessness and haste. The more complex and interesting the schedule and the greater the variety of respondents, the more interviews can be accomplished before 'instrument decay' sets in. If a researcher lacks interviewing experience, the degree of detail will be limited, and questions need to be straightforward. In such a case there is a great deal to be said for non-interviewing techniques. With a single interviewer, or a small team relative to the number of respondents, a long time could elapse between the first and last interviews of a survey, giving possibly unwelcome seasonal effects, or those interviewed earlier might, in a small area, talk to those to be interviewed later, perhaps discouraging them from participating or 'polluting' their replies. Commercial surveys seldom allow more than 20 home-based interviews per interviewer with one schedule, and it might be sensible to consider forty or fifty as an upper limit. This means that for a survey by a lone researcher to produce quantifiable data, one of the self-administered methods will be necessary.

(vi) Resources

Where the subject matter and the population to be surveyed does not indicate whether interviewing or questionnaire methods are more appropriate, the resources available to the research project may be decisive.

Interviewing is usually considered expensive, since interviewers need to be paid for the time they spend travelling, calling unsuccessfully, and making appointments, as well as the time actually conducting interviews. If field-workers are students or members of a research team, however, the only costs to be considered may be fares, which for local surveys may not be large. However, an inexperienced interviewer is likely to bias the answers, and will probably meet with more refusals than a more experienced one. If training cannot be given, and time allowed for experimentation, it might be preferable to contact the respondents with a questionnaire.

The costs of postage and stationery for a postal survey could be prohibitive. Although both questionnaires and interview schedules need typing, reproduction and checking, for self-administered questionnaires the secretarial and clerical time needed to prepare, post or deliver, check and send out reminders can be considerable, and the quality of reproduction must be high.

Any research project is the outcome of a number of decisions, some of which are a compromise between the optimal and the feasible, given the available skills and other resources. Alternative methods should be costed roughly, and having made an initial selection on the basis of this chapter and Table 2, the researcher should read the chapters detailing the procedures in order to be sure that the methods finally adopted are likely to be those most successful in obtaining a good response and a sufficient quantity of analysable data.

TABLE 2 - MAIN FEATURES OF FOUR METHODS OF ASKING QUESTIONS

	Unstructured interviews	Standardised interview schedule	Delivery and collection questionnaire	Postal questionnaire
Kind of data collected	qualitative or exploratory	quantifiable	quantifiable	quantifiable
Suitable population	any	any	literate individuals who can be addressed by name; households or joint bodies such as firms	
Suitable questions	open-ended questions, rating scales, measures of attitudes, questions using visual aids or requesting special tasks (such as map drawing or identification of objects e.g. places on maps or photographs) can be used		open-ended questions can only be used in limited numbers; questions and tasks cannot be too taxing but should appear varied on the form	
Possible length	can be very long	depends on location	short - average completion time 5-30 minutes	
Sample size	not suited to a probability sample	depends on size of the field force, but cannot be too dispersed		large dispersed samples can be used
Response rate	not applicable	high - probably at least 70 per cent	moderately high	very variable
Time to complete fieldwork	depends on sample size and number of field-workers available			4-8 weeks from first posting
Main costs to consider	interviewers and travel costs; minimal amounts of stationery (plus tape recorders etc. if used)		field-workers and travel costs; stationery	postage and extra stationery; clerical time
Confidence in getting right respondent	high, but type and quality of information obtained will vary greatly	high	low, but can be checked at collection (although this will take longer)	low
Approximate number of respondents who can be contacted by one field-worker	30-40	30-50	200-300 (but it takes time to check questionnaires)	300-500 (but a single researcher could probably not code and analyse the response from more than 200-250)

III SELF-ADMINISTERED QUESTIONNAIRES

(i) Encouraging response

Having decided to use a self-administered questionnaire, the main difficulty is the possibility of a low response rate, and the procedures for using this technique are therefore designed to persuade the respondent to fill in the form. Questionnaires must be short, taking no more than half an hour to complete - and preferably less, but they must also contain interesting questions and seem important to the respondent. They must be attractively presented and clearly understood; great attention should be paid to layout (section VI, iii).

A letter of introduction attached to the front of the questionnaire should be used even when the form is to be personally delivered, so that the respondent cannot separate it from the questionnaire and will be reminded of the importance of the survey at the time he considers filling it in (appendix 1).

There is evidence that respondents are not fooled by gimmicks, gifts, fancy coloured paper or being entered into competitions. A handwritten signature, a handwritten postscript or an authoritative body sponsoring the project may influence some respondents, but the basic differences in response are caused by the contents of the letter itself, and how valuable the respondents are persuaded that the survey is (Scott, 1961, p.173). The letter should state the aim, importance and sponsorship of the research. Being honest with the respondent may be more fruitful than leaving open to doubt the commercial links of the enquiry. Assurances of anonymity or confidentiality are important, and respondents seem to accept their genuineness even when they are asked for their names and addresses, which are needed to avoid sending unnecessary reminders.

(ii) Postal surveys

The majority of replies will generally be received within a few days of posting out the forms, but those who fail to reply at once may be very different from the immediate respondents. It is therefore usual to send out reminders to those who have not replied after a short time has elapsed, in order to obtain replies from them.

The first questionnaire is sent with a return envelope, and is usually followed up with a reminder letter and finally another reminder letter, a duplicate questionnaire and return envelope, giving a maximum of three contacts. The interval between each stage is commonly seven days, but it can be more flexible and depend on the flow of replies. It is tempting to leave it a little longer, to reduce the number of reminders needed, but the greatest gains are to be made from the follow up when it is received before the questionnaire has been forgotten or mislaid. If it is to be left longer than a week, a second copy of the questionnaire should probably be sent with the first reminder.

The seven day interval has the added attraction that many people will only find time to complete questionnaires at the weekend; each stage of the process can be posted on the same day of the week (Wednesday or Thursday) in the hope that they will arrive on Friday or Saturday. Replies cannot be expected in any great numbers after 4 or 5 weeks from the first posting, and the researcher will have to decide when to call a halt and commence coding and analysis.

The tone of the accompanying letters can be varied, and appendix 1 gives two examples. Robin (1965) suggests that the first should ensure confidentiality and emphasise the importance of the research project and its aims, but that later letters should emphasise the importance for the representativeness of the survey that the individual himself replies. It could recognise that the most interested are likely to have replied already, and the letter should therefore be directed particularly at those who are busy, or reluctant, emphasising that they are needed to make the survey comprehensive. If the majority of questions are only relevant to a part of the sample and it is suspected that those who have replied first are those who have simply had to write 'not applicable', the letter could be a special request to those who have to spend a little longer on the task, apologising for taking up their time, and stressing how important it is that they take the trouble to reply.

If the respondent is being asked to return the form through the post, clearly he will have to be supplied with a stamped envelope large enough to contain it. It has been suggested that 'business reply paid' envelopes, such as some institutions can supply without having to make a special application to the post office for a licence, are less personal, and the respondent feels less guilt about wasting them than he does with a stamped envelope. Nevertheless, reluctant respondents will have to be sent two or three stamped envelopes, and this can greatly increase the cost.

Clerical procedures for handling a postal survey need to be worked out in advance. Forms are usually serial numbered, and checked to ensure that the number corresponds to the correct address. An automatic numbering stamp will facilitate this process and will be faster than putting the address on the form as well as on the envelope.

For small surveys, a check on replies can be made from a carefully ordered master list. As each questionnaire comes in, it should be dated (so that early respondents can be compared with later ones), checked to see that the reply is usable, and that the respondent has not removed the address or serial number. It is then ticked off on the master list. This necessitates copying all the addresses not yet marked on to envelopes at each reminder stage.

For larger surveys, this constant copying is too time-consuming, and it is usual to type four labels for each address at the outset, for which special carbon labels can be obtained. Scott (1961, p.191) suggests the most economical way of organising reminders once labels have been prepared in this way. Instead of having a master list, the control consists of a set of envelopes, one for each address; one label is stuck on the outside, the rest are put inside. Each label should contain the address and the serial number.

Using the control envelope method, the first label is sent on the initial posting of the questionnaire, and the second and third are put in the envelope. When a reply comes in, the control envelope for that address is removed from the box. When the time comes to send out the first reminder, another label is taken out of each envelope still in the control box, and similarly, at the time the final reminder is sent out, the last label would be taken out and used. It would be possible to use only three typed labels if the control envelope itself were used for the final reminder, but then a master list would be needed to check the final response. The control envelope method uses more stationery, but saves time over a large number of addresses.

No would-be postal surveyor should under-estimate the clerical work involved in either the master list or the control envelope method; both are tedious and need to be done carefully. The advantages of a postal questionnaire are greatest when a large number of forms are to be sent out, but procedures which seemed so simple for a pilot stage become enormous once the whole sample is involved. Even filling the envelopes can take a very long time.

In estimating the supply of envelopes and stamps needed, as well as the number of follow-up letters and questionnaires to produce, the researcher will have to estimate his likely response rate at each stage. Unfortunately this varies greatly from one survey to another.

If similar surveys had produced response rates between 75 and 80 per cent, a rule of thumb might be to assume that each stage will produce a third as many replies as the preceding one, so we could expect 54% in the first stage, 18% in the second, and 6 per cent in the final stage. Thus we would need stationery to send out forms and reply envelopes to 100% in the first stage, 46% in the second, and 28% in the third. But these estimates might be wrong for our particular survey subject and population, and flexibility in the production of stationery for the later stages, for example xeroxing or duplicating rather than printing, is desirable.

(iii) Delivery and collection

Postal surveys may be the only way to study a dispersed population, but where the study area is compact, or a clustered sample design with a few small areas can be used, it may be preferable to use the method of delivery and collection by hand. A researcher who lacks clerical and field-work help, such as an undergraduate conducting a project, may find this the best alternative, since it enables him or her to make contact with enough respondents to produce quantifiable data without involving the costs of postage.

The field work stage can be shorter than a postal survey, although this will depend on how many addresses are assigned to each field-worker.

Delivery enables some immediate checking to be made; addresses which are empty or demolished can be eliminated. Additional information can be obtained about the area. A decision will have to be made about what the field-worker does if he receives no reply; if he puts the form through the letter-box, he should mark it accordingly so that such cases can be analysed separately, at least for response rate. There is some evidence that personal delivery, even where personal contact is not made at delivery, is an encouragement to reply.

Collection enables additional encouragement to be given, and if the respondent needs reassurance about the survey's credentials, or is unsure about how to fill in the form, the field-worker can help him. If this is done the respondent will have been, in effect, interviewed, and his responses may differ from those of respondents who have administered the questionnaire to themselves. The researcher should be prepared for this eventuality, and have instructions about what wordings and paraphrases are permissible, and how much probing can be permitted. Much of the chapter (IV) on interviewing will apply.

Supplementary information about the accommodation can be collected by the field-worker, and he could also administer additional questions, perhaps for attitudinal data, or of a more complex structure, but this would tend to reduce response as he could only collect the form from the chosen respondent and not another member of the household. The method is also suitable where a sub-sample of respondents is going to be asked for more detailed information, or to fill in diaries, as the personal contact involved in the screening process might increase co-operation. Additionally where the respondent refuses to fill in the questionnaire supplementary information gives the researcher some idea of the characteristics of the non-respondents.

Walker (1976) suggests that a single field-worker could drop 200 forms over two days, preferably Wednesday and Thursday evenings, and collect them over the next weekend. Alternatively it would be feasible to deliver one weekend and collect the next, to maximise the chance of making personal contact and to minimise the time in which the respondent could lose or forget the form. If no contact is made at the collection visit, the researcher has to choose between call-backs and delivering a duplicate form with a stamped addressed envelope in the hope of obtaining a postal reply. The delivery and collection method is suitable for piloting postal surveys, because the method is much quicker and would enable the respondent to be asked about any problems he had encountered in filling in the form, although it could not estimate the response rate to be expected by post.

Because the response rate is likely to be higher than for a postal survey, the delivery and collection method might be preferred where the area to be covered is small and travel costs will be low.

(iv) Screening with self-administered questionnaires

When it is necessary to have a small amount of information about a large number of individuals, households or businesses, and then to find out more details of a sub-sample, self-administered techniques can be used. A common situation is screening a population for a sub-group for whom no list could be obtained from which to sample.

The response rate achieved will be combined with the response rate at the second stage, and if the initial response is low, a very unrepresentative sample could result. It is probable, however, that at the interview stage the refusal rate would be lower as respondents would feel that they had already begun to participate in the research project even if they had not specifically agreed to an interview, and screening questionnaires will tend to be short, and thus to achieve a high response rate (bibliography, section E).

The delivery and collection method is particularly appropriate for screening, as, if the relevant sub-sample could be readily identified from their forms, the field-worker collecting them could conduct the interview at once or make an appointment for a more convenient time.

IV INTERVIEWING

(i) The conduct of interviews

While on the surface an interview should appear to be a conversation between interviewer and respondent, it should be a carefully controlled outcome of a long process of design and preparation. The interviewer must establish rapport with the respondent, but this must not be at the expense of standardisation and structure. All questions, however factual and straightforward, must be asked clearly and uniformly to all respondents; there is no point in conducting a survey in which the sample has been selected with sufficient care to ensure that the whole population which the study purports to concern is in fact represented if each member of the sample is then asked a different set of questions, or questions which he can interpret idiosyncratically.

The interviewer will have a code of practice to ensure consistency, particularly in the case of probes and alternative wordings which can be used. Otherwise, on the spur of the moment, a well-meaning interviewer may rephrase or elaborate a question in such a way as to alter its meaning or suggest an answer. Large surveys will give interviewers general instructions and specific notes on each question; even a small study should make these notes, which can be produced during the training and piloting period. Examples are given in Atkinson (1971) and Young and Willmott (1973, pp 302-316).

Probing (Section VI, v) requires skill, in knowing when an answer is insufficient and could be expanded, and how to obtain a more detailed answer without prejudicing the response. A 'don't know' to a factual question may require a probe, but in all cases where this has been necessary, the interviewer should record this as many respondents will resent being pressed - or may be using the 'don't know' to conceal a reluctance to discuss something they consider personal. Where a list of alternative answers to attitude questions have been pre-coded on the form to facilitate the interviewer's task of recording, it should be made clear if they are to be read to the respondent or if they are to be applied to the answer which he has given in his own words. If they are the latter, and are read out (or worse still, if part of the list is read) the response will have been biased. There will be a tendency for both interviewer and respondent to apply one of the categories given, even if none is applicable.

On rare occasions a respondent will attempt to confuse or falsify; generally people try to help or please the interviewer. This leads to the possibility that nodding or other 'reinforcement' may bias the responses, because a respondent may attempt to impress and provide the answers that appear to be desired; the interviewer must present as neutral an approach as possible.

Interviewing is an exacting and lonely task. It can be very depressing, particularly if specific individuals are being sought and repeated call-backs have to be made. A decision will be needed about the minimum number of calls that must be made (although a field-worker can use discretion about more if there is a good chance of an interview). Certainly three or four attempts at different times of day will be needed, and space will be needed on the front page of the schedule for recording the calls (appendix 2).

Normally, large surveys will have provision for supervision and the discussion of problems, but a research team or group of students working together can do quite a lot to help themselves through the field-work by checking each other's interviews for omissions or discrepancies as soon after completion as possible, so that the original interviewer can sometimes remedy the deficiency. Each interviewer should make a report on each interview, notes which will help the coding stage, or provide life for the final report; it is also a check on accuracy and can relieve the boredom of an evening's rigid application of a structured schedule.

Where there are several interviewers it is desirable to check consistent differences between the answers they have obtained, at the analysis stage, although when the sample has been allocated to interviewers on the basis of areas, as will usually be the case, it can be extremely difficult to distinguish interviewer bias from actual differences in the population studied.

(ii) Interview length

Interviews can take place in a variety of locations, usually the street, home or workplace. The North American telephone interview would probably meet with great suspicion in Britain. Where the interview is conducted will affect the length and complexity, as well as the sensitivity of the topics which can be covered.

Many geographical surveys would be of very factual information, and privacy would not generally be sought, but it is worth noting anyone else present, as they may exert an influence (either a positive one, reminding the respondent of facts, or negative, discouraging him from reporting things he has done, or things he considers personal but might have confided to the impersonal interviewer).

Interviews should have a maximum duration of about ten minutes in the street, and 20 to 45 minutes in the home, but this will vary with the respondents' likely interest in the subject matter. Some schedules are so long they fatigue the respondent, or severely disrupt his day. Separate short interview surveys with good response rates may be more useful than a single survey with a high percentage of incomplete interviews.

The researcher will need to calculate how long field-work will last on the basis of how many interviews can be conducted in a day, which is a function of average duration plus the time taken to make contact, as well as of the hours of work. Most studies in the home can only be conducted from 5 to 9 p.m., or at the weekend.

Paring down an interview schedule will force the researcher to justify every question in terms of his research hypotheses. It will however be important to maintain flow in the interview, and 'superfluous' questions, or statements explaining abrupt breaks in the subject of the questions, may be used to do this. The interviewer's skill in keeping the respondent from ranging too far from the topic will also prevent the interview from lasting too long. If offered tea or asked his own opinion, he should be able to suggest tactfully that it would be better to get the questions out of the way first.

If attitudes are being studied, he can explain frankly that he does not want to influence the respondent as it is each individual's opinions that are being sought.

(iii) Starting the interview

The interviewer will need to have some form of identification which he can show readily to reassure the respondent and dissociate himself from advertising campaigns. It may be best to have a letter to leave with the respondent, on the headed paper of the department or sponsoring body, briefly explaining the main concern of the survey and giving the name of the interviewer, as well as thanking the respondent for participating (appendix 1). This letter will enable the respondent to see who is responsible for the project, so that he could contact someone responsible for the survey if he had any queries. Experience indicates that this will rarely happen, but it is important to make this information freely available to the respondent. The letter should be brief, as the respondent may well read it through before agreeing to the survey; it should only give the outline of the project, which can be amplified on request.

Larger surveys or studies of institutions may find local press publicity or a letter sent to respondents beforehand to be useful, but smaller projects will usually have to make all their explanations direct to the respondent.

Interviewing in the street, where the interviewer has to approach people who fulfil specified criteria (for example, they are the nth person to leave a shop or other building), will require a clearly defined and brief statement of the aims of the survey which can encourage the person approached to stop. He should be told how much of his time will be taken up, what the aims of the survey are, and who is sponsoring it.

Where the field-worker has the name and address of an individual, the approach is different. Whoever answers the door may need some explanation for why the named individual is sought, what the survey is about, and how long the interview will take. The introductory remarks should be concise and firm; hesitation will invite refusal. The interviewer will have to establish that he is talking to the correct person, the one named, or one fulfilling a specification (such as the main house-keeper); some sample designs will require the deletion of addresses where the named individual has moved, and others will require the new occupant to be substituted (Dixon and Leach, 1978).

It may be helpful to explain the concept of representativeness and random selection to justify talking to that person rather than someone else.

(iv) Interviewers

In general, people who feel strongly about topics are not suitable interviewers, even when the questions appear to be factual. Students who undertake studies of subjects that concern them, for example, traffic in an area they consider congested, might introduce subtle biases. It is also hard for a researcher to interview on his own survey, because new ideas which develop during the interviews may become incorporated, or he may start to make redefinitions. For these reasons, as well as the problem of the limited number of interviews that one person can conduct, it is usual for interview surveys to use a team of field-workers. Pilot interviews can be done by the researcher, but the final pilot must be by the real interviewing team.

Academic research often uses students as interviewers, but in general students and academics make poor interviewers, both in terms of accuracy and obtaining responses. Married middle class women between 25 and 45 are almost always the best interviewers, and professional survey institutions invariably refuse to employ students. Understanding why this is so will enable some of the problems to be overcome whenever, in the course of individual or group projects, students must carry out interviewing.

Students may get lower response rates because of their age group, which inspires less confidence, and because they are often male, when women, expected to be more sympathetic listeners, get higher response rates. They should therefore pay more attention to their appearance and manner. Suspicion of interviewers is natural; wearing dark glasses, for example, can only increase this.

The professional interviewer makes it clear that she is just doing her job, and her business-like approach makes the interview seem matter-of-fact. The student should try to present the same professionalism, confident that this is just part of his course, and avoid undue hesitance. He should also, whoever he is called upon to interview, present a neutral and uncritical approach.

Interviewers need to be thorough, accurate and sufficiently educated to understand the purpose of the survey and the need for consistency and accuracy. Very lively people tend to become bored more easily, inconsistent and inaccurate.

If a researcher lacks the resources to use good interviewers, he might consider whether the self-administered methods would not be more appropriate, as poor response rates and highly variable quality will invalidate a survey, however good its interview schedule.

(v) Training and preparation

There are probably some people who will never make good interviewers, but training can overcome some of the problems. This does not have to be lengthy, but it is necessary to be thoroughly acquainted with the schedule being used as well as with techniques of gaining co-operation and retaining interest (Atkinson, 1971).

Most of the literature assumes that the training process is of a group of professionals, but many of the ideas contained can be adapted to a self-training situation for a small group of inexperienced researchers. Listening to others interviewing, practice and criticism, dummy interviews, or tape recording one's own interviewing, can help with shyness and other problems. Members of a group realise that their feelings of diffidence are common. Ultimately the only way to learn is to carry out trial interviews (this can also be the time at which question wordings are tested), either with supervision or in pairs, so that problems encountered can be discussed.

V ASKING QUESTIONS

(i) Questions and answers

There is little point in trying to collect information that respondents are unlikely to have, or can give only in a partial or inaccurate form. If pressed, most will give an answer to satisfy the interviewer, and research has shown that many people will express knowledge of or opinions about entirely fictitious things if they are asked about them in such a way as to imply that they should have heard of them.

Asking one member of a household about others, asking detailed questions about finances, or asking about complex facts which the individual either never knew or has forgotten will only irritate him. While space must be left for genuine 'don't know' answers respondents will feel justifiably aggrieved if asked a series of questions to which they do not know the answers.

It is often helpful to let the informant know at the outset how much is required of him; he will want to know how much of his time will be taken up, and what sort of information he is being asked to provide. With tact, people can be persuaded to look things up, for example in their accounts.

A recurrent problem in survey work is to know how far the researcher can rely on respondents telling the truth. Much will depend on the subject of the enquiry and the purpose to which it will be put, as well as the respondents' image of it. The interviewer or the explanatory letter will have to explain clearly why the questions are being asked and give assurances of confidentiality. Clearly great care with wordings and layout, as well as willingness to explain the survey, will help.

There is also a problem when respondents answer truthfully, but have modified their behaviour. In a survey about recent activities, at least one respondent, forewarned of the date of the interview, had been on a mammoth 'pub crawl' in order to have something lively to report. The problem also arises with 'diaries' (section V, iv) where the respondent is asked to keep a record of his behaviour or what has happened to him. A postal questionnaire should not ask about what was done 'yesterday' since the respondent could choose which day he was going to fill in the form, and would, in fact, be most likely to do so on a restful day rather than a more active one, biasing the sample's 'yesterdays' in favour of days when respondents were too busy to complete the form.

Questionnaires and interview schedules should be pruned carefully to ensure that every part will contribute to testing the hypotheses. Goode and Hatt (1952, pp 149-151) suggest preparing dummy tables to show what kinds of cross-tabulations could be obtained from asking specific questions. While few researchers probably follow this advice, it is helpful to consider how many categories are going to be needed. For example if a small sample were divided into a large number of groups for analysis, each would be too small for meaningful statements to be made. In general, data should be collected in a form as close as possible to that in which they are to be used. If it is known that the population will be divided into broad categories, and their limits can be determined in advance, these should be used as pre-coded categories on the questionnaire or schedule. The only exception would be if small sub-groups were to be combined in several different ways, or the researcher could not be sure in advance what categories would be most suitable.

(ii) Defining key variables

One can think of a number of distorting influences in answers to 'How many pints of beer have you drunk in the last seven days?', but problems can also arise in asking, for example, 'How many fruit trees do you have?'. Here the problem is almost certainly one of definition, for if we include old trees no longer bearing much fruit, or young trees not yet productive, we should get a different answer from that if we only count trees producing a crop. Some respondents might ask for clarification, and the interviewers would be unlikely to produce a standard definition on the spot.

Definition is the main problem associated with collecting apparently straightforward factual material. As far as possible, definitions adopted should be those of official statistics, or correspond to those used in other studies, so that data collected are comparable. Advice on key topics, such as housing and education, is given in several of the references in the bibliography section G.

It is not always possible to foresee all the problems of definition, and some will be found by piloting. Where a group of researchers are working together, daily discussion sessions during the pilot stage will be invaluable.

Geographers will encounter particular difficulties with definitions of such spatial terms as 'locality', 'neighbourhood', 'region' and 'area'. Some perception studies may seek to ascertain exactly what respondents understand by these terms, but in other projects the assumption is made that there will be some consensus of meaning attached to them. A clarifying definition appropriate to the study will then be needed. Using maps to define such an area, though, may merely test experience with maps, and should be supplemented by a description.

Asking about income is not only potentially difficult because it is a sensitive topic, and the answer could vary with wishful thinking or the image of the interviewer and the body he represents, but because of the need for definition; should the question ask about net or gross income, should it include overtime payments, or family allowance?

Even assuming co-operation could be obtained to give full details of finance, in order to obtain accurate figures computations as detailed as the tax form would be needed, which would be quite impossible. The majority of surveys of the general population are going to use a limited number of income categories to divide up the sample, and great precision would be superfluous. However, in particular studies of sections of the population, additional or miscellaneous income might form a large proportion of the family budget, and it might be reasonable to ask for more detail.

The British Government's Social Survey Division include the following in their definition of income: all earnings, overtime payments, bonuses and tips, profits and salary from business, all kinds of pensions, sickness and unemployment benefits, family allowances, private means from investment or property, and payments from non-members of the household (Atkinson, 1971, p.130

Income data are best collected by presenting categories given in weekly, monthly and annual units so that the respondent does not have to convert his income from the terms in which he knows it most readily.

A decision will also be needed about whether income is to be collected for an individual or for the household; the latter will present difficulties in households with several wage-earners, but in the former case, there may be household income difficult to attribute to one person, such as rent from a lodger.

Many projects avoid asking about income, regarding it as too sensitive and likely to prejudice the rest of the questions, as well as prone to unreliable reporting. It is, however, a major determinant of many of the things geographers will be studying, and if the research concerns expenditure, shopping or many aspects of behaviour or perception, some income information will be invaluable. Having made a suitable definition, the section on wording embarrassing questions (VI, ii) should be studied, bearing in mind the fact that, if the respondent can see the necessity of asking such a question, he can probably be persuaded to reply. He should be told that the survey needs to relate the information collected to the types of people interviewed, emphasising that it is the types rather than the individual names that are important.

Occupations too, are frequently collected as classifying information, to divide the population into groups. The Registrar General (1970) publishes periodically a list of detailed job descriptions which enable workers to be divided into the 'social classes' and 'socio-economic groups' used in the Census. The information needed to make use of this reference work is the name of the job, plus, for problem cases, the number of people employed, the number supervised, any training or qualifications, whether self-employed, and the industry. 'Cleaner', 'engineer', 'manager', 'nurse' or 'machinist' are by no means self-explanatory, and fuller details will be needed. Although the reference book is to be used at the coding stage, and the form should record the details provided by the respondent, it will be helpful to look at this book before embarking on interviewing, to learn when to probe for more detail.

Since some respondents have jobs whose natures are clear from their titles, for example, bus conductors, a single question on occupation is impossible to word in advance. The interviewer will ask 'What is your (the head of your household's) job?' and will have an instruction beside it 'Obtain full details of occupation and industry', and be allowed to probe freely until confident that sufficient information has been obtained. These probes should, like all permissible wordings devised by the interviewer on the spot, be neutral, not suggesting any particular job or status, but asking 'Could you give me more detail?', 'Can you explain your job a little more?' or 'What do you actually do?'.

On a postal questionnaire, there will be no interviewer to probe, and the question must suggest to the respondent the necessary degree of detail. 'Please describe the work you do in as much detail as you can. If you are self-employed, please say so. If you employ other people, please say how many'. In a questionnaire which is to be collected, this will almost certainly be a question to check at the time of collection. If it is placed on the last page and easily identifiable, the field-worker can glance at it and ensure that it is adequately answered.

The researcher will have to decide whom to define as unemployed; many people not actively looking for jobs might work periodically. Similarly, some people may describe themselves as 'retired' or 'housewives' when they are in fact scanning the newspapers for jobs every day. Work is usually divided into full-time and part-time; a dividing line will be needed, for example the 22 hours which qualify for redundancy payments. More detail than this would not normally be necessary.

Individuals are frequently classified by their sex and age. Depending on the study, age may be conveniently collected in decades, in 'age last birthday', or for young people occasionally the date of birth may be collected in order to convert the age into years and months at the coding stage. Marital status is also often relevant, but it is usually not the legal status that matters and common-law partnerships would be classified with marriages. Educational qualifications may well be sought for some surveys; the age at which education finished or qualifications obtained may be used. Atkinson (1971, pp 135-137) lists the most commonly encountered qualifications and education information collected, and shows which qualifications can be considered comparable; if more detail is required, Stacey (1969) should be consulted.

The household will often be the unit of investigation, or alternatively information will be needed about it in addition to data directly about the respondent. One definition of household is the group of people who usually live together at the same address, not necessarily related to each other, who share catering arrangements or eat together. Some research projects will want to include spouses working away from home, seamen, children away at school and college, and so on, others may want to exclude all not in the house at present. Atkinson (1971, p. 114) suggests asking 'How many people live here regularly, who are catered for by the same person as yourself?', but the form of the question would depend on the definition chosen.

It is usual to list all the members of a household in a box, which allows for age, sex and perhaps occupational or marital status details as well. One simple example which could be modified to provide more details if they were necessary for the survey would be:

Tick Respondent	Relationship to head of household	Sex		Age last birthday	MARITAL STATUS		
		Male	Female		Married	Single	Widowed Divorced Separated

If the household details were to be collected from an individual in a postal questionnaire, then it might be simpler to ask him to give the information in terms of relationships to himself. This would avoid presenting him with a definition of household head - a concept which has a variety of interpretations but usually means the person who owns or rents the address occupied by the household. Some surveys will also define a 'housewife'; this might be the person responsible for domestic duties, irrespective of marital status, or indeed of sex, but other studies might want to confine their attention to married women responsible for a household's domestic arrangements.

Finally, information is often wanted about the accommodation. The types in an area will have to be noted down for the pre-coded categories; some surveys will want to distinguish between high rise and other types of flats, others between detached, semi-detached, terraced houses and flats. An estimate of size can also be obtained, usually by asking the number of rooms (defined appropriately) or the number of bedrooms. The Census asks about 'living rooms, bedrooms and kitchens', and excludes bathrooms, toilets, storerooms and kitchens below a minimum size.

Similar attention to definitions will be needed where data are being collected about units other than households or individuals. Information about industrial or agricultural concerns would normally want to be comparable with government statistics, and these should be consulted before wording questions. For example activities of firms would probably be related to the standard industrial classification, which comprises 27 main industrial groups each of which is further sub-divided into descriptions of main products or processes (Central Statistical Office, 1968). Land use maps and the Agricultural Census would similarly provide appropriate bases for comparison with land use, livestock and farm labour force data from the survey.

Frequencies present difficulties, for example in studies of behaviour. It is common to want to know how often people do certain things, but if people are asked their 'normal' behaviour, the result often considerably over-estimates the frequency. For example, motorists might say that they had their cars serviced at regular intervals, but in practice would occasionally delay or omit a service.

It is therefore not adequate to ask 'How often do you buy anything in the market?' because the answer will almost certainly be an approximation, probably an over-estimate, and, if behaviour is erratic, may be related to contingent phenomena ('I only go when my husband is home in time to look after the children').

How then can we establish the actual frequencies with which individuals do whatever we are interested in? The answer, as always, depends on what we want to know. It might be appropriate to ask: 'How many times in the last week ...', but then the unit of time will have to be defined, for example: 'I would like to ask you about your shopping in the last seven days, that is, from last (the same day as the interview) to yesterday. How many times in the last seven days ...'

Alternatively, the question could be tackled by asking 'When did you last ...' and then 'And when was the time before?'. This avoids defining a time period, and copes with behaviour which might not have occurred in the last week. It also removes the problem of 'transference' of events from outside into the chosen period of reference. The method is, however, unable to deal with irregularity. If we were asking about participation in sport, we might happen to interview someone who played football on Saturday and Sunday, and went to training one night in the week. Our estimate of the frequency with which he played would then depend on which day we interviewed him.

It might therefore be preferable to obtain an estimate of 'normal' occurrence by asking 'How often ...', and then how many times in a given period of reference. The disadvantage here is that if a large number of items are to be asked about, for example shopping at a different location or for a variety of specific goods, this method would be very tedious.

(iii) Memory

There are many occasions when a piece of research will be asking about events, circumstances or behaviour in the past, rather than about present conditions. The memory, though, is not an automatic reckoner, and can be highly inaccurate. Memory decay is rapid but uneven; exceptional events are remembered for some considerable time, but often incompletely.

Memorable events that geographers might study would include moving house; we should expect respondents to be able to recall dates, prices and other features of moves even if they occurred some years before. They could also be encouraged to remember by asking them to relate them to things which could be dated such as the ages of their children at the time of the move.

Careful structuring of questions will reduce the tiring effect of remembering, but the respondent may feel he is being tested. We might be interested in perception of disasters, and ask 'We are interested in what items of news have made an impression on people. Do you happen to remember what took place in Aberfan in 1966?'. There is always the danger that the respondent will feel this is a 'test', but piloting alternative wordings may suggest one that is suitable; when we probe to find out how clear the memory is, we could ask 'Do you remember any of the details of it now?'

The other type of recalled data sought is about a short recent time period, as discussed in the preceding section. The most appropriate time period would be one over which memory decay was not too pronounced, and yet within which the behaviour in question was reasonably likely to have occurred. Food eaten or television programmes watched might be forgotten rapidly, and it might be necessary to ask only about the preceding day. Each day's interviews would have to form a separate sample, because if all the non-working housewives were interviewed in the week, and working people at the weekend, biased estimates of the population's behaviour might be calculated.

A more usual time period for data of concern to geographers is a week or a fortnight. In studies of young people's activities outside the home conducted by one of the authors, the same number of events was reported in each week of a two-week period, suggesting that over this period the effect of losses due to bad memory was constant, but we cannot know if certain types of event were forgotten altogether.

Gray (1955, p. 360) suggests checking time period questions with dependent questions, which might help with memory problems. For example, if a respondent reported that he had seen the doctor in the week, he would be asked the reason, and if he said that he had not, he would be asked when he had last been to the doctor. A respondent would not be tempted to falsify (for example, by including a visit eight days ago, or leaving out one six days ago) since each initial reply would entail an additional question. On a self-administered questionnaire, this could be quite important in convincing each respondent that the form applied to him.

(iv) Diary techniques

Willcox (1965) was able to compare his morbidity data with medical records and concluded that a diary technique, where the respondents filled in illnesses as they occurred over a period of a few weeks, was more reliable than a retrospective interview.

The use of diaries is really beyond the scope of this booklet, but some observations will indicate the circumstances in which they will prove useful.

The problems of persuading respondents to co-operate, and ensuring that they maintain the record, are considerable. In successful examples, respondents were captive populations, such as school children who completed the diaries for the previous day as part of their school work, or payments were made. A posted request would probably meet with little success, even if the task itself were not arduous.

A major problem is that a respondent, knowing that he has the diary to complete, may modify his behaviour or act more consciously than usual.

Setting out a suitable form is also difficult; if the diary concerns journeys, how many should be allowed for each day? Allocating too little space will lead to omissions, too much will be discouraging. If times are used, the use of half hour intervals has been shown to produce a large number of activities with thirty minutes' duration; an hourly interval produces activities which last an hour. One solution is to give several key times with a few blank lines between (for example, Young and Willmott, 1973, pp 353-360).

Analysing the material may be troublesome, as information has to be extracted in a comparable form. The number of journeys for particular purposes could be counted, or the length of time spent on each activity - but in many cases could also have been obtained from a lengthy interview. Burton (1971) discusses approaches to analysing diary material.

In other studies, diaries have been used as case studies to illustrate and give life to a formal survey analysed quantitatively (for example, Willmott, 1967) and it is this use which might be most appropriate unless the resources of a project permit slow and expensive analysis.

VI DESIGNING QUESTIONNAIRES OR SCHEDULES

(i) Question flow

After the interviewer has explained the objects of the research, or in a questionnaire, after the explanatory letter, initial questions should be straightforward and relaxing. They should be clearly related to the stated aims of the research. More complex questions or sensitive issues should be left until later, but since fatigue or boredom will set in, they should not be postponed too long.

It is usual to ask 'classifying' questions, age, occupation, and so on, last. The respondent will need to be told that the researcher wants to know these details to see if people in different situations have different opinions or experience. Appendix 4 gives an example of the sort of classifying information that might be sought, and the way of introducing it.

Asking for personal details first, rather than presenting the more interesting questions which form the substance of the study, would tend to discourage respondents. It is only done when information is needed to establish if questions are relevant to a particular respondent, or with certain sample designs. This may happen when addresses are used to obtain a sample of individuals in a specific category (such as housewives), or with 'quota' samples, in which a specified number of respondents is needed in each of several categories, often defined by criteria like age and occupation (Dixon and Leach, 1978).

Questions should flow in a logical sequence, and where the subject matter necessitates sections, some explanation of the switch to a new topic may be necessary, either on a questionnaire or in a statement which the interviewer reads out.

Many questions will apply to only some respondents, and it is both irritating and time-wasting to ask irrelevant questions. Where possible, questions will be skipped after a 'filter' question has established they are inapplicable. For example, in a study of holidays, the question 'Where did you go for your holiday last summer?' presumes that the respondent went somewhere; if he did not, he might be aggrieved to have his behaviour condemned by implication as abnormal. A filter in a self-administered questionnaire might take the form shown in Figure 1(a).

Figure 1(b) illustrates the use of filtering in an interview schedule, using the example of the hypothetical survey of pensioners' shopping mentioned earlier - a survey, incidentally, that might encounter severe interviewing problems, but certainly could not be carried out with a questionnaire.

Figure 1: EXAMPLES OF FILTER QUESTIONS

(a) A self-administered questionnaire

9. Did you go anywhere outside the British Isles on holiday last year (1977)?

YES ☐
NO ☐ SKIP TO QUESTION 14

10. Which country or countries did you visit?

....

11. What means of transport did you use to leave this country? (If you went on more than one holiday abroad last year, please use one column for each).

	Holiday 1	Holiday 2	Holiday 3
AIR			
FERRY			
HOVERCRAFT			
HYDROFOIL			
OTHER (please write in)	....	....	....

(b) An interview schedule

			CODE	
6.	ASK ALL			
	Do you go out to do any of your own shopping for groceries?	YES, DOES SOME	1	
		NO, NEVER	2	→Q18
7.	IF DOES SOME SHOPPING			
	Do you use the shop on the estate	frequently	1	
	(RUNNING PROMPT)	occasionally	2	
		never	3	→Q10

As well as to avoid asking inapplicable questions, filtering can be used to reduce tedious repetition. The following extract from a farm survey is needlessly repetitive:

> How many dairy cattle did you have last year?
> How many dairy cattle do you have now?
> How many beef cattle did you have last year?
> How many beef cattle do you have now?
> etc.

A number of alternative shorter and less jarring formats could be devised, for example, asking the present number of cattle, and what type they are, followed by a filter asking if numbers have changed over the year (which should be defined, probably as 'this time last year') asking the magnitude of change only if farmers have reported that their livestock numbers are different.

(ii) Wording

One of the essential features of standardised surveys is that each respondent should be asked the same set of questions, presented in the same order. It is therefore crucial to ensure that wordings chosen are comprehensible, and that they express exactly what the researcher intended to all respondents. Most questions could be expressed in a number of ways, and although some alternatives will be better than others, there is not usually a single correct solution to a wording problem; what follows here are merely guidelines.

Simple, short questions which avoid technical or vague terms are preferable. However, what is short and precise to the researcher may prove very different to some or all his sample. The jargon that social scientists use is by no means always understood. Goode and Hatt (1952, p. 133) suggest answers to 'What is your marital status?' of 'Fine', 'As good as can be expected', and 'I'll have to ask the wife'.

Ambiguity is usually minimised by keeping questions short, and a series of short questions will be better than one long one, or one which really contains two questions ('How far have you come and how long did it take you to get here?') even if the result might sound abrupt. It will be easier to record answers, and ensure that all questions have been answered.

There are regional, class and age differences in the use of many words. Schofield (1965) discusses the use of the vernacular, and its regional variations, in a study of the sexual behaviour of young people. In such instances, the researcher would probably discover any confusion all too soon, but in the surveys that geographers undertake, questions which have been misunderstood may go undetected. Colloquial expressions are generally used, but the interviewer needs to know what to say if the respondent clearly misunderstands.

It will be helpful at the piloting stage to ask a group of people what they think each question means, rather than simply asking for replies.

Many terms are open to a wide variety of interpretation. 'Holiday' was used in a previous section, but this could include time off work, day trips, staying with relatives, or several weeks away from home. (The British Tourist Authority defines a holiday as an absence from home of four nights or more).

Clearly the researcher must decide a definition which he wants to apply. It can be helpful to give the definition first in order to avoid overloading the question. 'How many people are there in your household, I mean, counting all the people who normally share meals with you and sleep here?' could be reformulated: 'I want to ask you about your household. By that I mean ...' In some cases it might also be helpful to repeat the definition in the question. In a self-administered questionnaire, short sentences are again preferable, and it might be sensible to separate the definition and the question.

It is possible to bias a question so as to lead the respondent to a particular answer. The question 'Do you like shopping in the new pedestrian precinct?' could be rephrased more neutrally, giving both positive and negative aspects equal weight, for example, 'Do you like or dislike shopping in the new pedestrian precinct?' It might be better to ask, at more length but with greater clarity, 'I would like to ask you about the new shopping precinct. Do you prefer shopping there now it is closed to traffic, or did you prefer shopping there before?' But this raises other problems about whether the individual can think in such terms; other changes will have taken place in the prices, displays, or type of shops, and the respondent may not be able to produce an answer about an overall 'preference' (see section VII, i).

The simplest form of biasing, though, is not in the wording at all, but in the coverage, for asking questions about traffic, and subsequently about problems in an area, could produce more responses mentioning traffic than would otherwise be the case. A respondent may subconsciously form the impression from the earlier topics that certain kinds of answers are sought, and give them.

Double negatives are to be avoided; in the question 'Do you agree or disagree with the view that the town centre should not be open to traffic on a Saturday?' the attempt is neutrality, and the result confusion, for a yes is a yes to not-to-open.

It is usually best to assume that the respondent will answer every question. Apologetic wording ('Would you mind telling me your weekly income?') may invite refusal or suggest that this is a topic which many people might refuse to answer.

Many researchers preface their questionnaire or interview with some variation on the statement: 'Some of the questions may appear to be rather personal, but this is because we want to find out what people think. We will, of course, not use names or any form of identification in our study, only statistical totals'. This implies that it would be normal to find the questions personal, but abnormal to want to deprive the research project of the answer, and in some studies this form might be appropriate. For factual studies such as most geographical research, it would be unnecessary and might lead the respondent to look for hidden meanings. In interviewing, any question which might cause embarrassment will be left until late in the proceedings when the interviewer has gained rapport and there is less likelihood of refusal to answer. The interviewer must, in spite of establishing an easy relationship with the respondent, remain sufficiently impersonal for embarrassing questions to be dealt with professionally. If there is a refusal to such a question, its position at the end of the interview will not prejudice later questions, and less will be lost if the interview terminates abruptly.

Barton's (1958) list of alternative ways of asking 'Did you kill your wife?' indicates a variety of possible ways of dealing with potential reluctance or embarrassment. Among these may be mentioned techniques which suggest the behaviour is normal ('As you know, many people have been killing their wives these days. Do you happen to have killed yours?') or the completely open 'Did you ever kill your wife?'

Another approach would be to deflect the question, by asking 'How did you kill your wife?', leaving it to the respondent to deny that the question is relevant to him. Normally we should not wish to antagonise people by asking them inapplicable questions, but in such a case it could be justified. Alternatively, a 'prompt card' containing categories can be presented. Each is given a letter or a number, so that an embarrassed respondent does not have to say 'I drowned her in the bath', but 'number four'.

This 'visual aid' is useful not only for sensitive topics, but also where a scale (agree strongly, agree, disagree, disagree strongly, for example) or classes have to be applied a number of times: if the respondent has it in front of him, the interviewer does not need to read it out each time. Not all respondents will have adequate eyesight and literacy, and the interviewer will have to be prepared to read out the categories without suggesting that ability to read is required. Some populations will preclude this method. Immigrants may not be literate in English, and the pensioners mentioned earlier could not be expected to use this method.

No matter how straightforward questions appear, there can be unforeseen difficulties. The question 'Could you tell me the number of cars owned by members of your household?' appears to present no problems, provided that the interviewer knows what to do about company cars, vans, and so on, but it produced in at least one case the registration numbers of the cars concerned, showing an ambiguity which had escaped the research designer. Such problems will only be discovered by pilot testing.

(iii) Layout: self-administered questionnaires

A well-printed and laid out questionnaire can do much to increase the response rate of a self-administered questionnaire. Low quality paper and a staple in one corner will be a false economy in terms of securing co-operation, and after the innumerable stages of processing and handling, the researcher may be left with a disintegrating pile of paper, on which some writing materials may be erased. A serial number on each page is useful as even the best bound questionnaire will sometimes fall apart.

A compromise will have to be reached between a cramped style with little left white, and a thick pile of paper off-puttingly long. Printing on both sides of the paper will help if the quality is adequate. Space must also be left for codes and notes added at the coding stage.

Instructions must be clear and unambiguous; most respondents will be accustomed to forms such as application forms on which all questions have to be answered. Some features of the layout of official forms could be adopted, but the frequent necessity to refer to other sheets of notes or instructions is certainly not recommended; questions should be self-contained.

Generally a combination of open and closed questions is preferable; in the former, the answer is written in, in the latter, comprehensive categories are given which need only be circled or ticked (Figure 1(a)). This greatly facilitates coding, and will also be the quickest for the respondent. However, if there is no space for him to make comments, he may be frustrated or bored, and a few open questions, providing space to write in a few words or a sentence, can be more stimulating without appearing to require an essay.

Questions with alternative answers may be ringed or ticked, but all possibilities must be accounted for. The coding numbers (such as might be punched onto cards for computer processing) may confuse the respondent. Boxes which can be ticked, like those shown in the example in Figure 1(a), can be neatest.

A simple error which often occurs is to ask a question on age which presents a box for under 25 and a box for over 25, or a question on frequencies in the form 1/2 days, 2/3 days; in both cases the boundary is ambiguous. Piloting with an open question should enable all likely responses to be included, but a space for 'other' should be given, so that the respondent can write in. Where a large number of responses fall into the 'other' category, it can reflect a lack of pilot work to discover all eventualities.

As each postal form is received, it should have additional information such as the date of receipt, any area code or similar details, written on the front, if necessary on a separate attached sheet. A collected form will probably have an additional sheet of comments filled in by the field-worker stapled to it after collection.

(iv) Layout: interview schedules

Even when interviewers are highly motivated and well-trained in the use of a particular schedule, it will help them to have all the necessary instructions on the same page. The interviewer has to perform many operations simultaneously, asking questions, recording responses, maintaining rapport and deciding if an answer is sufficient or if a probe is necessary. He may have to interview standing up.

Any tools which make interviewing easier (clipboards, attached pens) should be considered. It will help to have to write on one side of the paper only, and spacing should ensure that questions are unlikely to be left out.

Space should also be left for marginal notes, where for some reason a respondent has refused to answer a question or sequence of questions, his answer does not fit into the pre-coded categories, or he seems to have misunderstood or be deliberately misleading.

The interviewer will also have to complete a classificatory page, which in addition to details such as the serial number, area code, date, day of the week, time and place of interview, the number of call-backs with the times and the days of the week, and the field-worker's name, may contain additional observations such as the type of accommodation. This would also be a suitable page on which to write general comments about the respondent and the conduct of the interview. Appendices 2 and 5 give examples.

Different type-faces should be used for those words which can be used to the respondent, and for instructions to the interviewer. In the example in Figure 1(b), and Appendices 3 and 4, capitals are used for instructions, and upper and lower case for questions to the respondent. The upper and lower case style is also used for the pre-coded categories in question 7, for these are part of the question, and are read to the respondent slowly as a 'running prompt'. This form of questioning is called 'forced choice' because the respondent is only allowed to answer in the terms of the set responses, since this is the only way in which the data can be analysed. It would be used for questions such as 'Are you married ...single...widowed or divorced?'. The interviewer would have instructions about people living together without being married, or those separated, who would be fitted into the categories given and not written in as 'other'. The 'forced choice' question is closed to the respondent, and the interviewer has only to ring or mark the appropriate code. Because a code is given to each answer in advance, analysis is made easier, and it is possible to punch computer cards straight from the form if the layout is clear (Appendices 4 and 5).

The forced choice question is only one type of closed question; it is also possible for the respondent to be allowed to answer freely, but for the interviewer to assign his reply to a category on the spot. Without extensive piloting, not all eventualities might have been allowed for in the original coding, and a space for 'other' would be left. Responses in this category would be listed separately as the interview schedules were completed, and new categories would be devised as necessary. These codes could then be written onto the schedules ready for the data processing stage.

Clearly, closed or forced choice questions, where all that is required is to ring a code, will make the interviewer's task easier, but interviewing also permits the use of more open questions, where the interviewer writes in the respondent's reply verbatim. These should not be used where the circumstances of the interview are likely to make writing difficult, for example in the street. They should not be used to save thought at the design stage or because piloting has been inadequate, nor to save space on the form, since at the coding stage it will be time-consuming to classify written answers which would have been pre-coded. Used sparingly, they can supplement the closed questions. For example, a respondent might be presented with a number of reasons for shopping in a particular centre (prices, variety, proximity, and so on) and then asked if there are any other reasons.

The whole schedule should be tested, particularly to discover tongue-twisters, places where the flow broke, or where the respondent did not understand. Different lengths, layouts and orders, as well as ways of explaining the survey, can be tried.

All completed schedules should be checked as soon as possible after completion to eliminate those severely incomplete and to clarify any ambiguities or illegible comments with the interviewer concerned. Spot checks, or preliminary tabulations of results, can reveal errors which are arising, for example through different interviewer practices.

(v) Probing

Deviations from the wordings given should not be permitted in interviews or in cases where a field-worker is delivering or collecting self-administered questionnaires and is asked for clarification. It follows from this that any definitions, explanations, prompts or probes will have to be determined in advance and known by all field-workers.

There are three situations when the interviewer will have to take decisions about appropriate wordings. In the first, as in the case of a question on occupation, or in the case of the 'household box' shown in section V, ii, he will have to devise questions appropriate to a particular case, using terminology discussed in advance. He will have to decide when he has obtained sufficient detail about a job or has been told all the household members. Where supplementary information is needed about a list of items, for example, in the case of the household box, where ages and other details are to be asked for each individual named, it is usual to complete the list first, before starting to ask the secondary information. This will minimise the possibility of omissions.

The second situation where the interviewer may be called on to use his own words is when it is apparent that the question, or a term used in it has not been understood. This circumstance should be minimised by the presentation of definitions to the respondent immediately prior to, or in the course of, each question. However, if the interviewer is called on to make an ad hoc definition, he should make a note of the difficulty in applying the original term to the respondent, and provide enough information for a factual question to enable the coder to apply the correct definition. Clarification from the research team should be immediately sought. In the case of opinion questions, explanations to the respondent could not fail to suggest answers to him, and none should be given. He should be told that the term means whatever he thinks it means (although care should have been taken in piloting to ensure that respondents generally would interpret a term in the same way). If a respondent failed to understand, he would have to be recorded as giving no response, and in any case the interviewer would make a marginal note about the confusion

It will be common for the interviewer to have to use discretion in applying probes to elicit a more detailed reply. Probes should be as neutral as possible. The simplest and most often used is a repeat of the original question, which the respondent may have misheard. There are also a variety of ways of asking for more information where that already obtained is inadequate or uncodeable. The interviewer could ask the respondent to repeat what he has said 'so that I can take it down', or ask 'could you explain a little more fully?' or 'in what way?' or 'is there anything else' as appropriate. Atkinson (1971, pp 72-83) discusses the use of these devices, and would be well worth reading prior to entering the field.

VII ASKING ABOUT ATTITUDES AND PERCEPTION

(i) Studying attitudes

In a study which is largely concerned with facts, it may be appropriate to ask respondents some opinion questions. Indeed, where an interview deals with a topic about which a respondent feels strongly, he may feel frustrated if he is unable to express his own views.

The attraction of studying perception of and attidues to a variety of environmental or planning phenomena has increased the need for specialised questioning techniques. Before asking about attitudes, geographers should consider the particular problems involved, and consult the basic texts (bibliography section M).

Attitudes are complex, and cannot exist in a vacuum. A researcher may ask a respondent to express an opionion about something he has never considered, or in an abstract, generalised way not natural to him. For example, most people could give an indication of their feelings about specific phenomena, such as each of the places in an area, but might not have generalised them to consider 'leisure provision' or 'planning decisions'. The respondent might answer a more general question, but this cannot be taken to represent his attitude before the interview. More than any kind of questioning asking about attitudes changes and crystallises the respondent's views. Sometimes we want the respondent to give a spontaneous reply, and at other times we can give him time to work out a considered opinion.

All that has been said previously about the care needed with deciding what information to collect, and wording questions unambiguously, applies more forcefully to attitude or perception studies. The necessity to pilot attitude questions more extensively than factual ones, and the pitfalls associated with discovering what a question is really measuring, preclude undergraduate projects successfully undertaking such studies.

The problem of definition arises in attitude studies as in factual questions, for if respondents are asked what they think of a place or an event, it will be important to establish that they are all referring to the same thing.

One way that geographers have attempted to find out attitudes has been to ask hypothetical questions. If we ask where someone would like to live 'if he had a choice', we have no way of knowing how freely he has made his imaginary choice. It may still reflect his financial, family and job situation. When a respondent has no control over the outcome of situations, to ask him his preferences is unrealistic.

Much of the perception literature concerns ranking places in terms of desirability, but it is unclear exactly to what extent it is knowledge, rather than preference, which is being measured.

Similarly, a respondent may be asked to choose between things he does not regard as comparable. Newspapers and television news fulfil different functions, and to ask which is preferred may be meaningless. Sophisticated techniques have been devised for studying choice. Respondents may be asked to trade off proximity to one feature against proximity to another, or to choose between them and features of the house itself, such as a garage or a bathroom. Or respondents may be asked to rank their choices, usually by giving them a series of cards and asking them to pick out the facility they would most like to be near, and then the one they care about least, and so on until all cards have been chosen. Before using any of these methods, the researcher must consider if they are actually likely to measure what he wants them to; in some cases questions have been inflicted on respondents which could not be answered meaningfully by the researcher himself.

Figure 2: EXAMPLES OF RATING SCALES

(a)

Over the last year, has traffic noise along this road:

- [ ] decreased a lot
- [ ] decreased a little
- [ ] stayed about the same
- [ ] increased a little
- [ ] increased a lot

(b)

'This street should be closed to traffic'.

DO YOU:

- [ ] disagree strongly
- [ ] disagree
- [ ] neither agree nor disagree or have no particular vi
- [ ] agree
- [ ] agree strongly

(PLEASE TICK ONE)

(c)

The list of statements which follows are some of the things which people say about traffic. I would like you to indicate how much you agree or disagree with them by placing a cross somewhere along the scale:

1. The traffic along this street is dangerous to children.

   Disagree |_____|_____|_____|_____|_____| Agree

2. Residents need to be able to park outside their own front door.

   Disagree |_____|_____|_____|_____|_____| Agree

(ii) Rating Scales

One of the commonest forms of attitude measurement has been to ask respondents to indicate their measure of agreement with a statement or a series of statements, or to choose from a list the one closest to their own view.

If an interviewer reads out a list, the respondent will be most likely to choose one from the beginning or the end. If the statements are printed on a questionnaire, those at the beginning may be preferred, particularly if they are the positive 'agree' part of the spectrum. It is usual, therefore, to place the negative categories at the top. Also it would be rare to test attitudes with one question, and a later one might well give a contrary statement, to see if the respondent is consistent when he has to disagree with what is presented to him.

There are innumerable ways of setting out rating scales; three examples are given in Figure 2. The first and second forms would be suitable for self-administered forms, but if used in an interview could be used in conjunction with a prompt card listing the possible responses, since they would probably be applied to several statements in a sequence. The third form would only be suitable where a field-worker was on hand to explain what was intended; it might form a self-completed part of an interview. A worked example would usually be shown to the respondent.

Where attitudes or perceptions are reported after a survey, it will be important to include the question or questions from which the attitude has been determined, for it has been demonstrated that different wordings can produce different conclusions. Before the referendum on the European Economic Community, one of the major opinion poll agencies found that they could produce substantially different percentages in favour of Britain's membership when they used different questions.

The problem is increased when more complex attitudes or perception are probed, or a phenomenon has different facets, but one attitude 'score' is wanted for each respondent. A researcher might want to derive overall scores for attitudes to places, let us say, desirability of holiday resorts. Respondents might be presented with a list of variables, and asked to give each town a score out of five for each aspect listed. The problem would arise because of the selection of attributes which might not necessarily be generally agreed to measure resort attractiveness. Adding the scores for, let us say, attractions of the surrounding countryside, amenities of the centre, and friendliness of the inhabitants, giving each of these three equal weight, would produce a curious effect. An individual might 'prefer' a place he has scored five on amenities and two on each of the other variables, to one to which he gave three for each, although both would obtain an overall score of nine. The problem increases in complexity as more variables are added, because the subjective judgment of the researcher determines how the index is constructed.

One solution is to have a panel of judges to select the items for inclusion, although this would involve almost as much effort as the survey itself, and another is to establish constructs, ideas, relevant for each individual, and present him with 'repertory grids' incorporating them; this is discussed in the next section.

(iii) Semantic differential scales and repertory grids

Semantic differential scales are a particular way of obtaining attitude information. The respondent is asked to relate particular ideas or constructs to named entities, such as places or people. For example, respondents might be asked about a succession of seaside beaches, with a sequence of polar constructs for each:

			does not apply				
stony							sandy
clean							polluted
popular							quiet
unsuitable for children							suitable for children

etc.

Not all these attributes are positive and negative; respondents might consider suitability for children good or bad. However, positive and negative points of the scale are not placed consistently at one end because this might bias the answers, particularly if a respondent had an overall view of a beach and did not bother to think individually about each construct.

The advantage of the technique is that, provided it does not go on too long, many respondents find it interesting. The problem comes with analysis, as for the rating scales. In the beach example, we could pick out those thought 'quiet' by a proportion of the respondents, or we could see that, perhaps, sandiness and suitability for children were highly correlated. Some of the variables would also be amenable to objective measure to compare with the subjective evaluations. What we would not be able to do, however, without a panel of judges, would be to give overall ratings to beaches.

Repertory grids are a technique devised to establish how individuals see the part of the world in which the researcher is interested in their own terms. In other words, while the grids presented to the respondents may resemble those of the semantic differential, the constructs used would be in some sense his own, and not those devised by the research team.

Appropriate constructs are entered on rating scales, or on grids where presence/absence or applicable/inapplicable are ticked, after an interview in which the respondent is asked to talk round a general field or investigation or is shown a series of pictures or words and asked to pick out the odd one in groups of three, stating why it is different. Although the constructs are those used by the respondent himself, it is the researcher who chooses which ones to use.

In many cases, the entities too are personalised, so that the individual would be asked to name beaches, rather than rating those given by the project.

The technique is time-consuming, and probably most conveniently performed on a volunteer or captive population, for example students or a school class. Analysis is difficult, as it is hard to generalise from the individual, with his personalised constructs and entities, to the group as a whole. The results are generally exploratory, but could be very stimulating in suggesting ideas other research might pursue. Oppenheim (1966, pp 204-211) introduces the technique, and Harrison and Sarre (1969) describe its application to environmental images.

(iv) Maps and photographs

For many projects, geographers may want to use maps, photographs or other non-verbal material. These visual techniques are discussed in the field of perception (Saarinen, 1976; Downs and Stea, 1973).

Some respondents find these items unfamiliar, and some have poor eyesight. Not everyone could identify their home on a street map, or home area on an outline of the British Isles. If places are to be identified, aerial photographs might sometimes be used. While they are less familiar than maps to most respondents, they have the advantage of being less abstract, particularly oblique views rather than vertical ones, and they have the added advantage of being sufficiently unusual to be interesting - which might assist in securing co-operation with what may be a time-consuming interview.

Asking respondents to draw maps can be particularly taxing, and even habitual map users may be reluctant to do so. We also cannot be sure that we have measured perception of the area in question, rather than drawing skill.

The point which has been made in other contexts, that most respondents will reply to questions, is applicable here: it would be beneficial to ask some at least of the respondents about the techniques, rather than simply using the visual methods and finding out if 'data' are obtained.

VIII SMALL SCALE PROJECTS AND ASKING QUESTIONS

No survey would be worthwhile without a substantial effort expended in testing the data collection instrument and methods. That is why pilot testing has been mentioned throughout the previous sections.

Possible methods and wordings should be tried on friends and colleagues, and then on a sample as like the final respondents as possible. These pilot interviews or questionnaires should be subjected to some preliminary analysis, and rough tabulations produced, so that they indicate how every stage of the main survey will proceed.

Both the small sample sizes and the length of time available for this pilot testing in the field reduce the validity of the research work conducted by students in geography courses. This is not to belittle such studies, however, merely to insist that greater thought is given to selecting manageable topics. Small projects can be a stimulating source of ideas, as well as enabling techniques of asking questions and obtaining answers to be learned in the only way in which they can be.

The desire to produce a finished, neatly tied up piece of research, while commendable, often leads to exclusion of details of data collection. Geographical research is an ongoing process, and one in which a number of workers participate. Others may be helped as much by the methods, problems encountered (and ways in which they were, or suggestions as to how they might be, overcome), as by the results of a survey. The questionnaire or interview schedule should be appended, together with any definitions or instructions. If consistent bias and data problems were experienced, they should be reported with the other findings, to enable even the smallest scale project to take its place in the research tradition.

BIBLIOGRAPHY

A. General texts on research design and asking questions

Davidson, J. (1970), *Outdoor recreation surveys.* (London: Countryside Commission).

Dixon, C. and Leach, B. (1978), *Sampling methods for geographical research.* CATMOG, 17, (Norwich: Geo Abstracts).

Goode, W. and Hatt, P. (1952), *Methods in social research.* (New York: McGraw-Hill).

Morton-Williams, J. (1972), Questionnaire design. in: *Consumer market research,* ed R. Worcester (London: McGraw-Hill).

Moser, C. and Kalton, G. (1971), *Survey methods in social investigatio.* (London: Heinemann).

Shepherd, J. (1978), *An introduction to survey analysis.* CATMOG (Norwich: Geo Abstracts).

SCPR, Social and Community Planning Research (1972b), *Questionnaire design manual.* (SCPR Technical Manual No. 5).

B. Unstructured techniques

SCPR, Social and Community Planning Research (1972a), *Depth interviews and group discussions.* (SCPR Technical Manual No. 4).

C. Postal questionnaires

Kaplan, S. and Cole, P. (1970), Factors affecting response to postal questionnaires. *British Journal of Preventive and Social Medicine,* 24, 245-247.

Robin, S. (1965), A procedure for securing returns to mail questionnaires. *Sociology and Social Research,* 50, 24-35.

SCPR, Social and Community Planning Research (1972c), *Postal survey methods.* (SCPR Technical Manual No. 1).

Scott, C. (1961), Research on mail surveys. *Journal of the Royal Statistical Society,* A, 124, 143-195.

D. Delivery and collection questionnaires

Walker, R. (1976), Social survey techniques: a note on the 'drop and collect' method. *Area,* 8(4), 284-288.

E. Screening with postal questionnaires

Gray, P. (1957), A sample survey with both a postal and an interview stage. *Applied Statistics,* 6, 139-153.

Hope, B. (no date), *Techniques used to identify a national sample of 14-20 year olds.* Government Social Survey, M148 (London: HMSO).

F. Interviewing

Atkinson, J. (1971), *A handbook for interviewers.* Office of Population Censuses and Surveys, M136 (London: HMSO).

Kahn, R. and Cannell, C. (1967), *The dynamics of interviewing.* (New York: Wiley).

Young, M. and Willmott, P. (1973), *The symmetrical family.* (London: Routledge and Kegan Paul), pp 302-335.

G. Reference Books

(a) British Statistical sources

Central Statistical Office (1968), *Standard industrial classification.* (London: HMSO).

Benjamin, B. (1970), *The Population Census,* (London: Heinemann).

Edwards, B. (1974), *Sources of social statistics.* (London: Heinemann).

Harvey, J. (1971), *Sources of statistics.* (London: Bingley).

Maunder, W. ed (1974), *Review of United Kingdom statistical sources.* (London: Heinemann).

Open University (1975), *Statistical sources.* (Milton Keynes: Open University). A series of 16 units, each dealing with a separate subject area, e.g. Unit 3: Population, Unit 4: Housing.

Pickett, K. (1974), *Sources of official data.* (London: Longmans).

(b) Definitions and ways of collecting specific data

Atkinson, J. (1971), *A handbook for interviewers.* Office of Population Censuses and Surveys, M136. (London: HMSO).
Defines classification variables used by the Social Survey Division in chapter II.

Benjamin, B. (1970), *The Population Census.* (London: Heinemann).
Lists definitions as well as describing the organisation of the Census.

Gittus, E. ed (1972), *Key variables in social research.* (London: Heinemann). Volume 1: religion, housing, locality; volume 2: race, politics, voluntary associations, stratification, demographic variables.

Office of Population Censuses and Surveys (1973), *The general household survey*. (London: HMSO).
Defines terms for households, income, qualifications and health.

Registrar General (1970), *Classification of occupations*. (London: HMSO).

Stacey, M. (1969), *Comparability in social research*. (London: Heinemann)
Sections on collecting data about education, family and household, income and occupation.

H. Memory problems

Cannell, C. (1973), The reliability of survey data. in: *Genetics and the epidemiology of rare diseases,* ed J. Neil et al. (United States Public Health Service Publication No. 1163).

Gray, P. (1955), The memory factor in social surveys. *Journal of the American Statistical Association,* 50, 344-363.

Willcox, K. (1963), A trial of methods for collecting household morbidity data. in: Neil, *op. cit.*

I. Diary Techniques

Burton, T. (1971), *Experiments in recreation research*. (London: Allen & Unwin).

Willmott, P. (1967), *Adolescent boys in East London*. (London: Routledge and Kegan Paul).

Young, M. and Willmott, P. (1973), *The symmetrical family*. (London: Routledge and Kegan Paul).

J. Question wording

The general texts in section A all deal with this problem; SCPR (1972b) and Goode and Hatt (1952) are particularly recommended.

K. Embarrassing questions

Barton, A. (1958), Asking the embarrassing question. *Public Opinion Quarterly,* 22, 67-68.

Schofield, M. (1965), *The sexual behaviour of young people*. (London: Longmans).

L. Layout

Examples of questionnaires and schedules are given in all the general texts in Section A, e.g. :

Davidson (1970), pp 57-77.

Morton-Williams (1972)

Moser and Kalton (1971), pp 306-339

SCPR (1972b), pp 86-87

Postal survey documents are given in:

SCPR (1972c), pp 45f

Scott (1961), pp 192-195

M. Attitudinal studies: introductions

Fishbein, M. ed (1967), *Readings in attitude theory and measurement.* (New York: Wiley).

Oppenheim, A. (1966), *Questionnaire design and attitude measurement.* (London: Heinemann).

SCPR, Social and Community Planning Research (1972b), *Questionnaire design manual,* 37-38.

N. Studies of attitudes and perception

Dale, P. (1971), Children's reactions to maps and aerial photographs. *Area,* 3(3), 170-177.

Downs, R. and Stea, D. (1973), *Image and environment.* (London: Edward Arnold).

Harrison, J. and Sarre, P. (1969), Personal construct theory in the measurement of environmental images. *Environment and Behaviour,* 1, 351-374.

Rowley, G. and Wilson, S. (1975), The analysis of housing and travel preferences: a gaming approach. *Environment and Behaviour,* 7(2), 171-177.

Saarinen, T. (1976), *Environmental planning: perception and behaviour* (Boston: Houghton-Mifflin).

Townsend, J. (1977), Perceived worlds of the colonists of tropical rainforest, Columbia. *Transactions of the Institute of British Geographers,* 2(4), 430-458.

APPENDIX 1: EXPLANATORY LETTERS

(a) A COVERING LETTER FOR A SELF-ADMINISTERED SURVEY OR AN INTRODUCTORY LETTER FOR AN INTERVIEWER TO GIVE RESPONDENTS.

University of __________
Department of Geography
(address)

(telephone number)

Dear Sir or Madam,

You have been chosen to take part in a survey of ____(main subject)____ which is being carried out in the Department of Geography, University of ______________________.

We do hope you will spare a little of your time to answer some questions about your own experiences and opinions. You have been chosen on a scientific basis from the Electoral Register, and it is important that everyone who is chosen does take part, so that we get a true cross-section.

SELF ADMINISTERED SURVEY ONLY { Please do not pass this form on to anyone else: it is your answers we need. We enclose a stamped envelope for your reply.

The information you give will be strictly confidential, and only statistical totals will be used in the study.

INTERVIEW SURVEY ONLY { Your interviewer is ______(name)______.

Thank you very much for your co-operation. If you have any queries or would like further information, please contact me at the above address.

Yours faithfully,

(b) A FOLLOW-UP LETTER FOR A POSTAL SURVEY WHICH USES A MORE PERSONAL APPROACH.

Dear ______________________

I wrote recently to ask you to take part in a survey of ___(main subject)___ which we are carrying out in the Department of Geography, University of ____________________.

By now many of the people I wrote to have replied, but I have not yet heard from you. I realise that the first letter may not have reached you, or that it may have been mislaid, so I am sending you another copy of the form.

We need to have your answers so that all experiences and opinions can be taken into account. Your reply will be kept strictly confidential, and will only be used for counting how many gave each answer; names will not be used.

I would be most grateful for your help.

Yours sincerely,

APPENDIX 2: FRONT SHEET FOR INTERVIEW SURVEY

Notes

The interviewer started out with the name and address of each respondent, serial numbers were added later for those successfully interviewed.

Information to be coded is placed conveniently at the right hand side; in this case, it includes some accommodation information, and categories appropriate to the survey area (in this example, a local authority housing development) are wanted.

Name .... Area code [ ]

Address .... Serial number [ ]

.... Sex

male 1

female 2

RECORD OF CALLS

	Date	Day of week	Time	Outcome and notes, e.g. best time to call back, who was spoken to.
1				
2				
3				
4				
5				

Accommodation type.

High rise flat ... 1

Low rise flat..... 2

Maisonette........ 3

House 4

Reasons for failure to secure interview:

[ ] moved

[ ] address vacant

[ ] address not found

[ ] not known at this address

[ ] refused

Private Garden

Yes 1

No 2

other ....

Interviewer's name .... Code no. [ ]

APPENDIX 3:

PART OF AN INTERVIEW SCHEDULE ANNOTATED AFTER PILOT TESTING

49. How did you get to the surgery on that occasion? (How would you get to the surgery, if you had to go?)

Remove hypothetical question. Can respondant predict behaviour in event of illness

What form of prompting should be allowed?

	CODE	POSITION ON PUNCH CARD
Walk all the way	0	
Bicycle	1	
Public transport	3	
Car driven by self	4	
Car driven by someone else	5	(41)
Taxi	6	
Some other way (specify) ------------------	7	
NA/NR	9	

(a) Do you (Would you) have any difficulties in getting to the surgery?

This hypothetical question is acceptable but should be asked separately (not coded with 'do you')

	CODE	POSITION ON PUNCH CARD
YES	1	
NO	0	
NA/NR	9	(42)

IF YES TO (a)

What is the main difficulty?

Pilot shows that these are not mutually exclusive — redesign

	CODE	POSITION ON PUNCH CARD
Difficulty crossing the roads	0	
Difficult to walk that far	1	
Have to depend on someone to drive me there	2	
Not on bus route	3	
Infrequent buses	4	(43)
Too expensive	5	
Other (specify) ------------------	7	
NA/NR	9	

(b) About how far would you say it is from your home to the surgery?

Why ask this? If we want distances (not perceived distances) ask the address and work it out

ESTIMATE IF NECESSARY

	CODE	POSITION ON PUNCH CARD
About $\frac{1}{4}$ mile or less	0	
About $\frac{1}{2}$ mile	1	
About 1 mile ($\frac{3}{4}$-$1\frac{1}{4}$)	2	
About 2 miles ($1\frac{1}{4}$-$2\frac{3}{4}$)	3	(44)
3 miles < 5 miles	4	
5 miles or more	5	
NA/NR	9	

(c) Would you say that the surgery was within reasonable walking distance for you, (if there was no other way of getting there)?

YES	1	
NO	0	(45)
NA/NR	9	

(d) Does anyone in your household have a car?

YES	1	
NO	0	
NA/NR	9	(46)

50. Approximately how long did it take you the last time (would it take you) to get to the surgery from home?

Remove hypothetical question — too many imponderables

CODE NUMBER OF MINUTES	---	(47-48)
DON'T KNOW	98	
NA/NR	99	

APPENDIX 4:

THE FINAL CLASSIFYING QUESTIONS FROM A STUDY OF GENERAL PRACTITIONER USE

Finally can I ask you a few questions about yourself, although of course you will remain anonymous in the final results:

102. Would you mind telling me who else lives in this household?

RELATIONSHIP TO RESPONDENT	NO. OF CHILDREN		TICK IF OVER 65	TICK IF WORKS
	< 5	5-15		

.? unemployed

APPENDIX 4 (continued)

MARITAL STATUS: ASK IF NOT APPARENT	CODE	POSITION ON PUNCH CARD
103. Would you mind telling me whether you are:		
not all categories necessary — Single	1	
Living with spouse/common law partner	2	
Divorced/Separated	3	(38)
Widowed	4	
No response	9	
104. Would you mind telling me what age group you are in according to this card?		
Under 24	1	
25 - 34	2	
35 - 44	3	
45 - 54	4	
55 - 64	5	(39)
65 - 74	6	
75 - 84	7	
85 or over	8	
No response ...	9	
105. Did you leave school as soon as you were old enough to do so?		
YES	1	
NO	0	(40)
NR	9	
IF NO		
(a) At what age did you leave school?		
CODE AGE IN YEARS	---	(41-42)
NO RESPONSE	99	
106. Have you had any further education or training?		
YES	1	
NO	0	(43)
NA/NR	9	
IF YES		
(a) What qualifications did you obtain?		
may be used as running prompt if necessary. — University/Polytechnic Degree/Higher Degree	1	
University diploma/Professional Qualification/Diploma in Art, Music or Teaching (Cert.)	2	

Technical College: HNC/HND, City & Guilds Full Certificate/Nursing qualifications..	3	(44)
'A' Level(s)/ONC/OND/City & Guilds 'Advanced' level	4	
Secretarial, Commercial (e.g. book-keeping)	5	
Apprenticeship (City & Guilds 'Basic' level)	6	
Other (specify) --	7	
NA/NR	9	

107. (a) Are you working at present?

		(45) self	(46) spouse
IF YES: RUNNING PROMPT	full time	1	1
	or part time	2	2
IF NO: Are you: RUNNING PROMPT	a housewife	3	3
	a student (full time)..	4	4
	unemployed	5	5
	or retired	6	6
	No response	9	9

(b) Is your wife/husband working at the moment?

108. (a) What is your present/was your last job? OBTAIN A FULL DESCRIPTION OF KIND OF WORK THIS JOB INVOLVES		*to be coded in office*
--		
--	---	(47-48)
define 'full time' --		
IF SELF EMPLOYED		
How many employees do you have?	---	(49)
NA/NR	9	
(b) What is your (ex) wife's/husband's present/last job? IF SINGLE AND LIVING WITH PARENTS ASK: What is your 'father's' present/last job?		
--		
--	---	(50-51)
--		
IF SELF EMPLOYED		
How many employees do you have?	---	(52)
NA/NR	9	

APPENDIX 4 (continued)

109. How long have you lived in this area? (Near enough to go to the same doctor if you wanted to).		
Less than 1 year	1	
1 < 2 years	2	
2 < 5 years	3	
5 < 10 years	4	(53)
10 < 20 years	5	
Over 20 years	6	
NR	9	
110. Do you rent or own your house (flat)?		
Owns/mortgage	1	
Rents from local authority	2	(54)
Rents privately unfurnished	3	
Rents privately furnished	4	
NA/NR	9	
Reword into separate question about council renting/ furnished/ unfurnished or running prompt		
111. Would you mind telling me who your doctor is? We will not tell him/her anything about what you have said to us, but we need to be able to find out a bit more about his/her practice.		
DR. ________________________________		
ADDRESS: ____________________________		
____________________________		

Thank you very much for your help

APPENDIX 5: THE FINAL PAGE OF THE GENERAL PRACTITIONER SURVEY (APPENDIX 4) TO BE FILLED IN BY THE INTERVIEWER

INTERVIEWER	---	(79-80)
INTERVIEWER NOTE:		
1. Respondent is: Male	1	(55)
Female	2	
2. Answer the following questions from responses to Q.104:		
(a) No. of adults in household (over 16 years)	---	(56)
(b) No. of working adults in household	---	(57)
(c) Total no. of persons in household	---	(58-59)
(d) No. of children living in household - under 5 years	---	(60)
5 - 15 years	---	(61)
(e) No. of persons (excluding respondent) aged 65 years or over in household	---	(62)
3. TYPE OF HOUSE		
Detached or large house	1	
Semi-detached, newer terraced house	2	
Flat - new or substantial	3	(63)
Older-style terraced house	4	
Council house	5	
Council flat	6	
Small flat, multi-occupied, bed sit	7	
4. BRIEF DESCRIPTION OF RESPONDENT		

---		
5. GENERAL PRACTITIONER CODES		
1. Total number of partners in practice	---	(64)
2. Number of surgeries that doctor does	---	(65)
3. Whether any surgeries are in a Health Centre: YES	1	(66)
NO	0	
4. Latest surgery time of that doctor	---	(67)